THE POETRY RADON

The Poetry Radon

Walter the Educator

Silent King Books

SILENT KING BOOKS

SKB

Copyright © 2024 by Walter the Educator

All rights reserved. No part of this book may be reproduced in any manner whatsoever without written permission except in the case of brief quotations embodied in critical articles and reviews.

First Printing, 2024

Disclaimer
This book is a literary work; poems are not about specific persons, locations, situations, and/or circumstances unless mentioned in a historical context. This book is for entertainment and informational purposes only. The author and publisher offer this information without warranties expressed or implied. No matter the grounds, neither the author nor the publisher will be accountable for any losses, injuries, or other damages caused by the reader's use of this book. The use of this book acknowledges an understanding and acceptance of this disclaimer.

"Earning a degree in chemistry changed my life!"
- Walter the Educator

dedicated to all the chemistry lovers, like myself, across the world

RADON

In subterranean depths, where shadows dance,

RADON

Resides a noble gas, in hidden trance.

RADON

Radon, the element of stealthy grace,

RADON

Invisible, yet leaving its haunting trace.

RADON

Born from uranium's decay, it emerges,

RADON

From darkened caverns, where silence surges.

RADON

A ghostly presence, ethereal and rare,

RADON

Invisible to the eye, yet ever there.

RADON

Its atomic dance, a spectral waltz,

RADON

Through soil and stone, it swiftly vaults.

RADON

Unseen, unheard, it creeps and crawls,

RADON

Through hidden channels, it silently calls.

RADON

In basements dank, it finds its lair,

RADON

Seeping through cracks, without a care.

RADON

Unnoticed by mortals, it lurks unseen,

RADON

A silent specter, in the in-between.

RADON

Yet danger dwells within its guise,

RADON

A toxic touch, a deadly prize.

RADON

For Radon's embrace, though veiled in mist,

RADON

Can bring to mortals a fatal twist.

RADON

With every breath, it infiltrates,

RADON

Into the lungs, it penetrates.

RADON

A silent killer, without a sound,

RADON

Its deadly grip, no mercy found.

RADON

But in its essence, there lies a tale,

RADON

Of nature's secrets, hidden and frail.

RADON

For Radon's presence, though feared and shunned,

RADON

Reveals the mysteries of Earth, begun.

RADON

In its decay, a story's told,

RADON

Of ancient rocks, in darkness cold.

RADON

Of elements forged in starry fire,

RADON

And buried deep, in Earth's desire.

RADON

So let us heed the silent call,

RADON

Of Radon's whisper, eerie and small.

RADON

For in its shadow, lies the key,

RADON

To understanding Earth's mystery.

RADON

In every atom, a universe unfolds,

RADON

In Radon's dance, the story molds.

RADON

Of creation's might, and nature's plan,

RADON

Invisible threads, connecting every span.

RADON

So though Radon may strike fear and dread,

RADON

Let us not forget, what lies ahead.

RADON

For in its quiet, there lies a tale,

RADON

Of Earth's secrets, in Radon's veil.

RADON

ABOUT THE CREATOR

Walter the Educator is one of the pseudonyms for Walter Anderson. Formally educated in Chemistry, Business, and Education, he is an educator, an author, a diverse entrepreneur, and he is the son of a disabled war veteran. "Walter the Educator" shares his time between educating and creating. He holds interests and owns several creative projects that entertain, enlighten, enhance, and educate, hoping to inspire and motivate you.

Follow, find new works, and stay up to date
with Walter the Educator™
at WaltertheEducator.com